SpringerBriefs in Computer Science

Series Editors

Stan Zdonik
Peng Ning
Shashi Shekhar
Jonathan Katz
Xindong Wu
Lakhmi C. Jain
David Padua
Xuemin Shen
Borko Furht
V. S. Subrahmanian

For further volumes:
http://www.springer.com/series/10028

Deepak Vohra

Rhodes Framework
for Android™ Platform
and BlackBerry®
Smartphones

 Springer

Deepak Vohra
e-mail: dvohra09@yahoo.com

ISSN 2191-5768   e-ISSN 2191-5776
ISBN 978-1-4614-3578-5   e-ISBN 978-1-4614-3579-2
DOI 10.1007/978-1-4614-3579-2
Springer New York Heidelberg Dordrecht London

Library of Congress Control Number: 2012934252

This book is an independent publication and is not affiliated with, nor has it been authorized, sponsored, or otherwise approved by Google or Research in Motion, and/or any of their subsidiaries.
References to various Google or Research in Motion copyrighted trademarks, characters, marks and registered marks owned by Google or Research in Motion and/or any of its subsidiaries may appear in this book. Rather than use a trademark symbol with every occurrence of a trademarked name, logo, or image we use the names, logos, and images only in an editorial fashion with no intention of infringement of the trademark.

© The Author(s) 2012
This work is subject to copyright. All rights are reserved by the Publisher, whether the whole or part of the material is concerned, specifically the rights of translation, reprinting, reuse of illustrations, recitation, broadcasting, reproduction on microfilms or in any other physical way, and transmission or information storage and retrieval, electronic adaptation, computer software, or by similar or dissimilar methodology now known or hereafter developed. Exempted from this legal reservation are brief excerpts in connection with reviews or scholarly analysis or material supplied specifically for the purpose of being entered and executed on a computer system, for exclusive use by the purchaser of the work. Duplication of this publication or parts thereof is permitted only under the provisions of the Copyright Law of the Publisher's location, in its current version, and permission for use must always be obtained from Springer. Permissions for use may be obtained through RightsLink at the Copyright Clearance Center. Violations are liable to prosecution under the respective Copyright Law.
The use of general descriptive names, registered names, trademarks, service marks, etc. in this publication does not imply, even in the absence of a specific statement, that such names are exempt from the relevant protective laws and regulations and therefore free for general use.
While the advice and information in this book are believed to be true and accurate at the date of publication, neither the authors nor the editors nor the publisher can accept any legal responsibility for any errors or omissions that may be made. The publisher makes no warranty, express or implied, with respect to the material contained herein.

Printed on acid-free paper

Springer is part of Springer Science+Business Media (www.springer.com)

Preface

Rhodes is an open source, Ruby-based, lightweight, MVC (model view controller) framework, optimized for mobile devices, which have memory limitations. The Rhodes framework offers several advantages over other mobile frameworks. Some of the unique features of the Rhodes frameworks are as follows.

- The only smartphone framework to offer support for the Model View Controller pattern.
- The only smartphone framework to offer support for the Object-Relational manager.
- The only smartphone framework to offer offline, disconnected access to data with the RhoSynch server.
- The only smartphone framework to support all mobile devices including Android™ platform smartphone, BlackBerry® smartphone, iPhone®, Symbian Platform, and Windows Mobile® operating system.
- Provides Ruby implementations for all smartphone device operating systems.
- Provides a web-based Integrated Development Environment for developing mobile applications for all smartphone platforms with the RhoHub development service.

Google's Android™ platform and RIM's BlackBerry® smartphone are the top two most commonly used smartphone platforms. Android™ platform has more than 40% of the smartphone market share. In Chap. 1 we discuss the Android™ platform. In Chap. 2 we discuss the Rhodes framework with the BlackBerry® smartphone. We develop the same Rhodes applications for Android™ platform and BlackBerry® smartphone; one application for a catalog and another for an RSS feed.

Contents

Chapter 1
Rhodes on Android™ Platform

Smartphones have proliferated in recent years creating a need for smartphone apps. Android™ is the most commonly used smartphone platform. Ruby is an open source, dynamic, interpreted programming language. Rhodes is the only framework for mobile devices that supports MVC architecture, and provides an Object Relational Manager. In this chapter we shall, first, introduce using Rhodes on Android, and subsequently develop a Rhodes application to get RSS feed for a magazine on Android. To parse XML Rhodes includes the RhoXML parser and support for the ReXML parser may be added.

1.1 Overview

This chapter which discusses the procedure to create a Rhodes applications for Android has the following sections.

- Installing the Android SDK
- Installing Ruby
- Installing Rhodes
- Creating a Rhodes Application
- Creating a Rhodes Model for a Catalog
- Creating a Rhodes Model to get RSS Feed

1.2 Installing the Android SDK

Download the Android SDK installer_r12-windows.exe from http://developer.android.com/sdk/index.html. Double-click on the .exe file. The Android SDK Tools Setup Wizard gets started. Click on **Next**. Android SDK requires Java SE Development Kit (JDK). Download and install the JDK, if not already

D. Vohra, *Rhodes Framework for Android*™ *Platform and BlackBerry*® *Smartphones*, SpringerBriefs in Computer Science, DOI: 10.1007/978-1-4614-3579-2_1,
© The Author(s) 2012

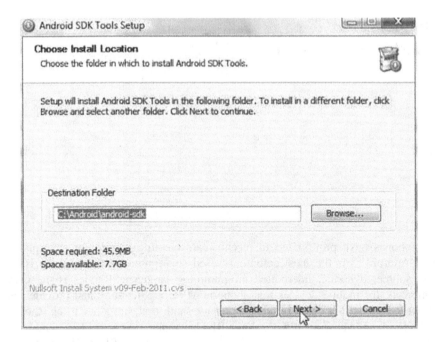

Fig. 1.1 Specifying the install location for Android SDK

installed, from http://www.oracle.com/technetwork/java/javase/downloads/index.
html. Install the JDK in a directory without spaces in the directory path. The Android
SDK Tools Setup wizard detects if the JDK is installed or not and displays a message
if the JDK is required to be installed. Click on **Next**. In **Choose Install Location**
specify the **Destination Folder**. Install Android SDK in a directory without spaces in
the directory path; for example not in the C:/Program Files/ sub-directory.
Specify Destination Folder as C\Android\android-sdk as shown in Fig. 1.1.
Click on **Next**. Click on **Install**.

Download Android NDK zip file from http://developer.android.com/sdk/ndk/
index.html and extract it to a directory without spaces in the directory path. Create
environment variables ANDROID_HOME for the Android SDK and ANDROID_NDK_
ROOT for the Android NDK. Add ANDROID_HOME/tools, JDK_HOME and
JDK_HOME/bin to the PATH environment variable. In a later section we shall
configure a Rhodes application to be used with the Android emulator.

1.3 Installing Ruby

Download the rubyinstaller-1.9.2-p180.exe application from http://
rubyinstaller.org/. Double-click on the .exe file to install Ruby. Install Ruby in a
directory without spaces in the directory path as shown in Fig. 1.2. Select the
checkbox "Add Ruby executables to your PATH". Click on **Install**.

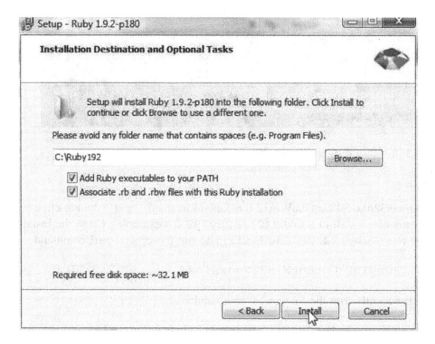

Fig. 1.2 Installing Ruby Installer

Next, install the Ruby Installer Development Kit (DevKit), which makes it easy to build native Ruby extensions. Download the `DevKit-tdm-32-4.5.1-20101214-1400-sfx.exe` application from http://rubyinstaller.org/downloads/. Double-click on the `.exe` file and install the self-extracting executables in a directory without spaces, such as `C:/Ruby192/DevKit`. Cd (change directory) to the `DevKit` directory and run the following commands:

```
rubydk.rbinit
rubydk.rb install
```

The output from running the commands is shown in Fig. 1.3.

The `init` command creates a `config.xml` file, which lists the RubyInstaller installed Rubies. Install RubyGems, a Ruby packaging system. Download the RubyGems zip file and extract the zip file to a directory, Cd to the directory and run the following command:

```
C:Ruby192\rubygems-1.6.2>ruby setup.rb
```

We also need to install `gnuwin32`, which provides win32 ports of GNU tools, GNU being a UNIX-like operating system. Download the `GetGnuWin32-0.6.3.exe` application from http://sourceforge.net/projects/getgnuwin32/files/

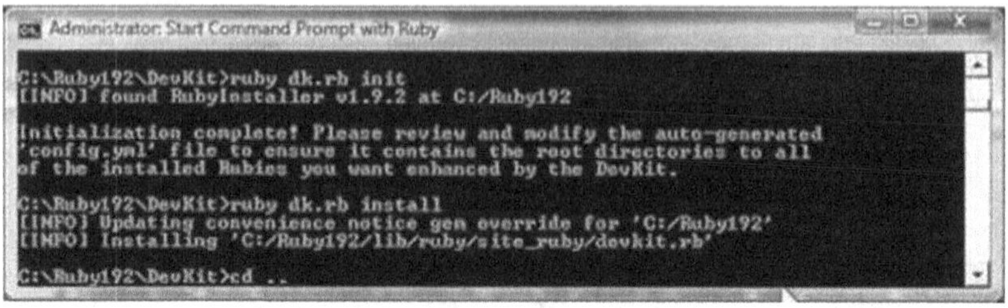

Fig. 1.3 Installing Ruby Installer DevKit

getgnuwin32/0.6.30/GetGnuWin32-0.6.3.exe/download and double-click on the .exe file. Install in a folder (C:/Ruby192 for example). Cd to the installed folder (C:/Ruby192/GetGnuWin32) and run the download command.

```
C:\Ruby192\GetGnuWin32>download
```

Subsequently, run the install command:

```
C:\Ruby192\GetGnuWin32>install C:/gnuwin32
```

Add C:/gnuwin32/bin to the PATH environment variable. Next, install Rake, a Ruby build program, with the following command:

```
C:\Ruby192>gem install rake
```

1.4 Installing Rhodes

Rhodes is a Ruby gem that is installed just like any other Ruby gem. To install Rhodes run the following command:

```
C:\Ruby192>gem installrhodes
```

Rhodes 2.3.2 and related gems get installed as shown in Fig. 1.4.
In the next section we shall create a Rhodes application.

1.5 Creating a Rhodes Application

Rhodes provides an application generator to generate an application. The Rhodes application generator is called 'rhodes' and is run with the following command format:

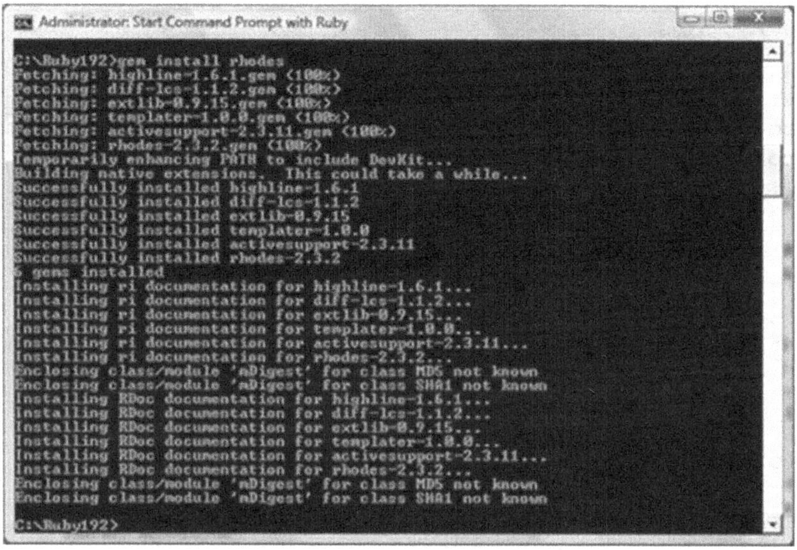

Fig. 1.4 Installing Rhodes

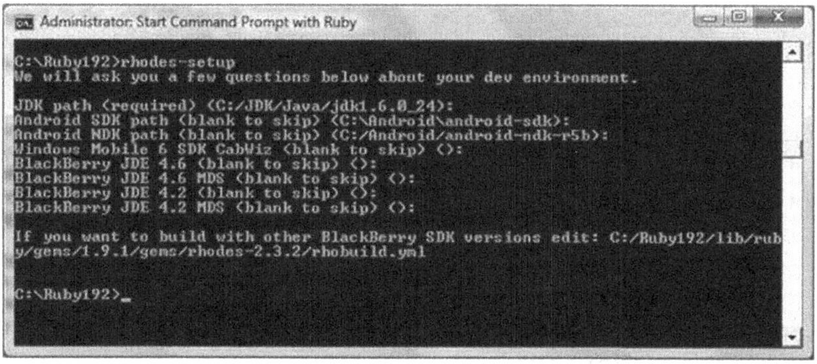

Fig. 1.5 Running the Rhodes-setup batch script

```
rhodes app <application_name>
```

Before we may run the `rhodes` command we need to setup Rhodes using the `rhodes-setup` command. Select Enter for each of the questions. The JDK path should not include any spaces in the directory path. The Android SDK path and NDK path should also not include spaces in the directory path. Cd to the `C:\Ruby192` folder and run the `rhodes-setup` command as shown in Fig. 1.5.

Modify the `C:\Ruby192\lib\ruby\gems\1.9.1\gems\rhodes-2.3.2\rhobuild.yml` configuration file, listed below, to include the Android paths. The JDK and Android paths are shown in bold.

```
env:
app: C:/rhodes-app
paths:
java: C:/JDK/Java/jdk1.6.0_24/bin
android: C:/Android/android-sdk
android-ndk: C:/Android/android-ndk-r5b
cabwiz:
    4.6:
jde:
mds:
sim: 9000
    4.2:
jde:
mds:
sim: 8100
build:
bbpath: platform/bb
wmpath: platform/wm
androidpath: platform/android
iphonepath: platform/iphone
symbianpath: platform/symbian
bb:
bbsignpwd: somepasswordhere
android:
excludedirs:
all:
  - "**/.*.swo"
  - "**/.*.swn"
  - "**/.DS_Store"
bb:
  - public/js/iui
  - public/js/jquery*
  - public/jqtouch*
  - public/js/prototype*
  - public/css/iphone*
  - public/iwebkit
  - public/themes
  - "**/jquery*.js"
    - "**/*.db"
```

The android parameter specifies the directory in which the Android SDK is
installed. The android-ndk parameter specifies the directory in which the
Android NDK is installed. Next, run the Rhodes application generator to create an
application called catalog with the command:

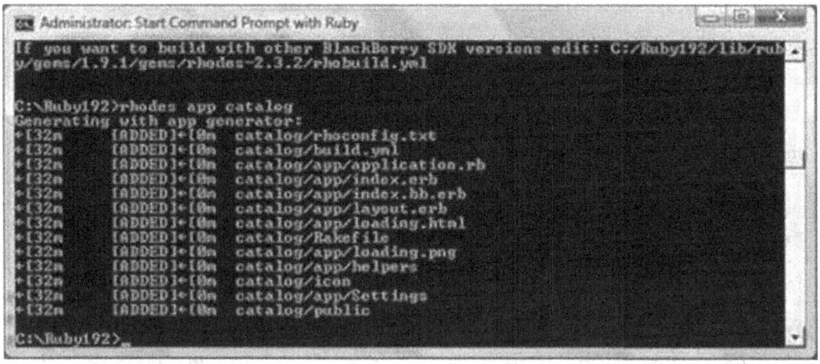

Fig. 1.6 Creating a Rhodes application

```
C:\Ruby192>rhodes app catalog
```

The application files get generated in the catalog (application name) directory as shown in Fig. 1.6.

The catalog/build.yml file lists the SDK install directory and the SDK version.

```
sdk: "C:/Ruby192/lib/ruby/gems/1.9.1/gems/rhodes-2.3.2"
sdkversion: 2.3.2
name: catalog
version: 1.0
vendor: rhomobile
build: debug
bbver: 4.6
wmsdk: "Windows Mobile 6 Professional SDK (ARMV4I)"
applog: rholog.txt
```

Before developing the application further test the Android emulator. Cd to the catalog directory and run the following command:

```
C:\Ruby192\catalog>rake run:android
```

The Android emulator gets started as shown in Fig. 1.7.

Select the default settings for an Android Virtual device, and click Enter when prompted with a question "Do you want to create a custom hardware profile?" as shown in Fig. 1.8. Only the first time the rake command is run the user is prompted.

The Rhodes application gets built to an Android application consisting of an Activity (RhodesActivity) and gets uploaded to the Android emulator. After loading is complete the Rhodes application gets started as shown in Fig. 1.9.

Fig. 1.7 Running the rake command to start an Android Virtual Device instance

The Rhodes application catalog is shown installed on the simulator as shown in Fig. 1.10.

Click on the `catalog` application to start the application. The catalog application loading starts as shown in Fig. 1.11.

Click on the default **Login** button as shown in Fig. 1.12.

Specify **Login** and **Password** and click on **Login** as shown in Fig. 1.13. The **Login** page is just a test login page, it does not really login into an application or website.

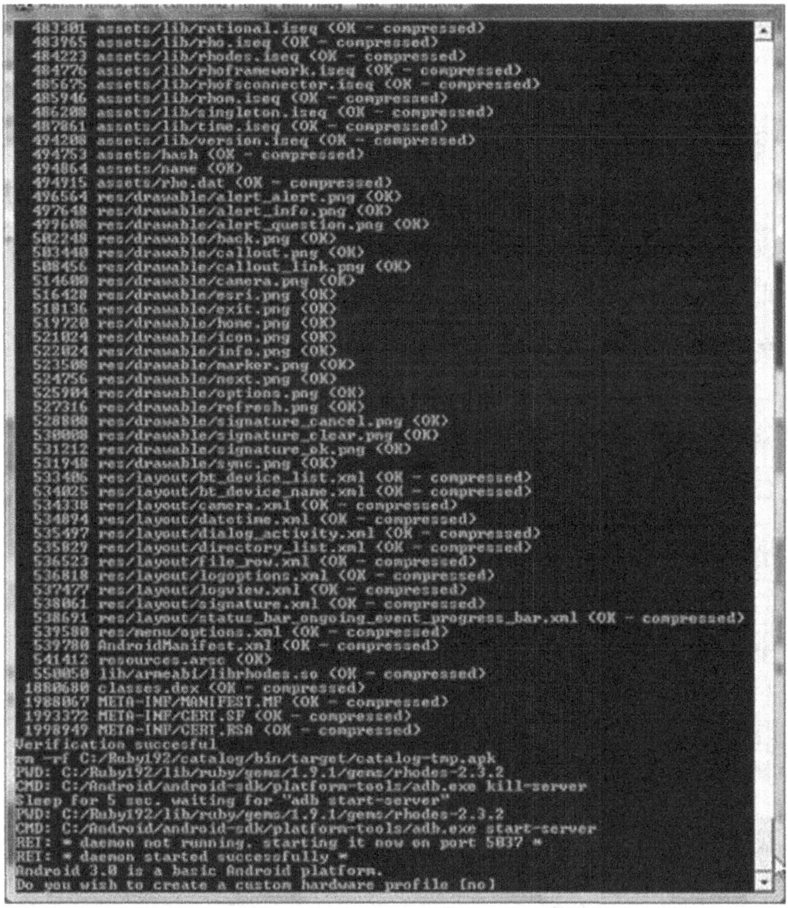

Fig. 1.8 Selecting the default configuration for an Android Virtual device

1.6 Creating a Rhodes Model for a Catalog

Rhodes provides the `rhodes model` command to generate model and controller files, and view templates, and is run with the following command format.

```
rhodes model modelname options
```

The `rhodes model` command generates a scaffolding similar to the Ruby on Rails framework to perform CRUD operations on the model. Cd to the application folder and generate a scaffolding for a catalog, which includes the `journal`, `publisher`, `edition`, `title`, `author` attributes.

Fig. 1.9 Uploading the Rhodes application and starting the activity

```
C:\Ruby192\catalog>rhodes    model    catalog    jour-
nal,publisher,edition,title,author
```

The view templates index.erb, edit.erb, new.erb and show.erb get generated. Controller file catalog_controller.rb and model file catalog.rb also get generated as shown in Fig. 1.14.

The controller class extends the Rho:RhoController class and includes actions index, edit, show, new, create, update and delete for CRUD operations.

```
classCatalogController< Rho::RhoController
end
```

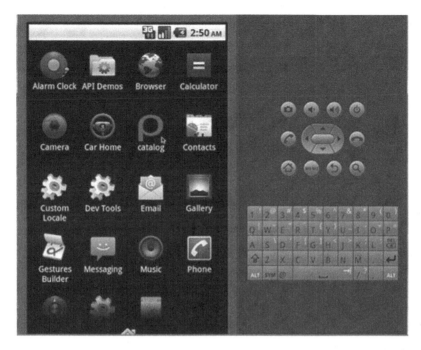

Fig. 1.10 Rhodes application installed in the Android Virtual Device

Fig. 1.11 Loading the Rhodes application

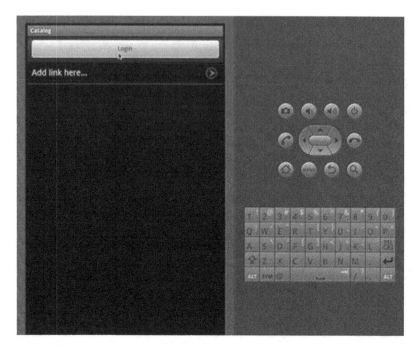

Fig. 1.12 Login button

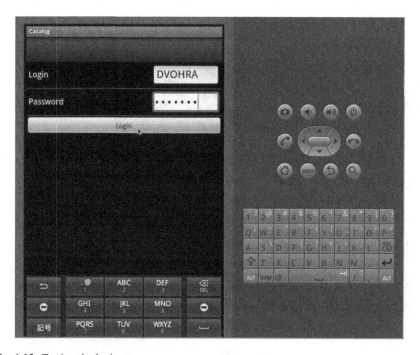

Fig. 1.13 Testing the login page

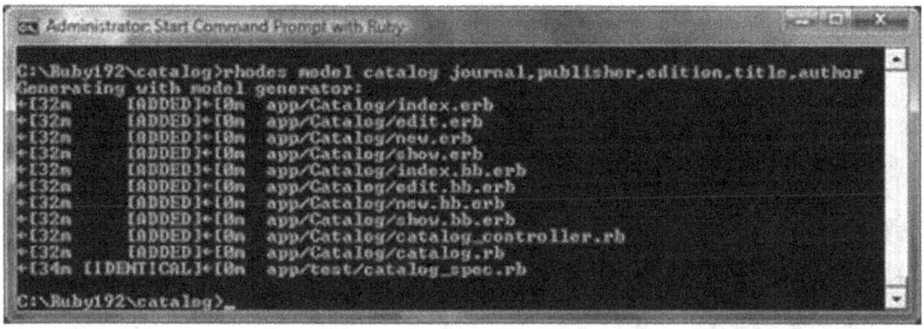

Fig. 1.14   Creating a model for a Rhodes application

Next, we shall upload the application to the Android emulator and test the application in the emulator. Before we may run the emulator we need to specify the application to run in the `catalog/rhoconfig.txt` file.

\# startup page for your application.

```
start_path = '/app/Catalog'
```

To run the emulator and upload the Rhodes model catalog run the command:

```
C:\Ruby192\catalog>rake run:android
```

The Android emulator gets started and the Rhodes application gets uploaded to the emulator. The catalog application gets started in the emulator. Click on **New** to create a catalog entry as shown in Fig. 1.15.

Specify **Journal**, **Publisher**, **Edition**, **Title** and **Author** and click on **Create** as shown in Fig. 1.16.

A new catalog entry gets created. Click on the icon for a catalog entry to display the entry as shown in Fig. 1.17.

The selected catalog entry's detail gets listed as shown in Fig. 1.18.

1.7 Creating a Rhodes Model to Get RSS Feed

In the previous sections we have only tested the default model generated by Rhodes. In this section we shall create a Rhodes model to get a RSS feed and display the feed in the Android. We shall use the IBM developerWorks RSS Feed (http://www.ibm.com/developerworks/views/opensource/rss/libraryview.jsp)   for the example. The RSS feed is in XML format and contains entries as `<item></item>` elements, the root element being `<rss></rss>`. Create a Rhodes model `CatalogRSSFeed` with attributes `title`, `link`, `description`, and `date` with the following command:

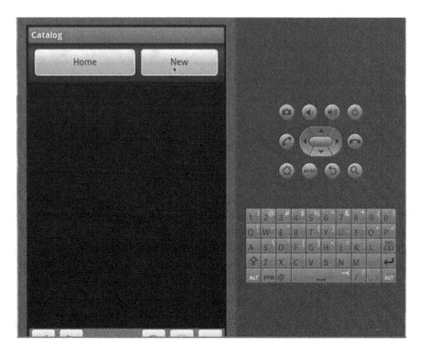

Fig. 1.15 Creating a new catalog entry

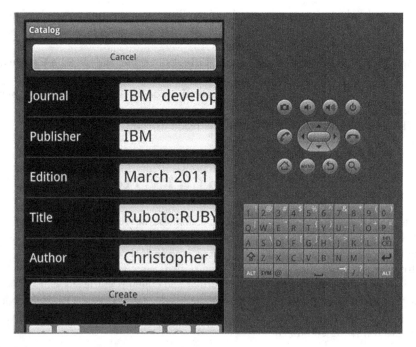

Fig. 1.16 Specifying fields for a catalog entry

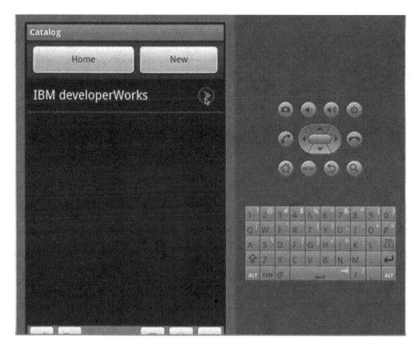

Fig. 1.17 A new catalog entry

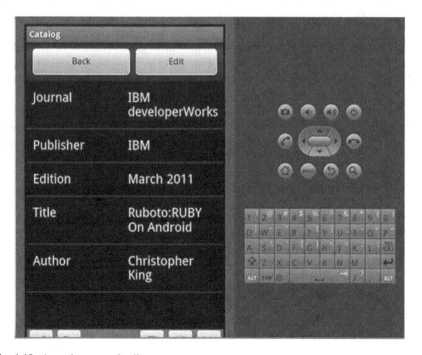

Fig. 1.18 A catalog entry detail

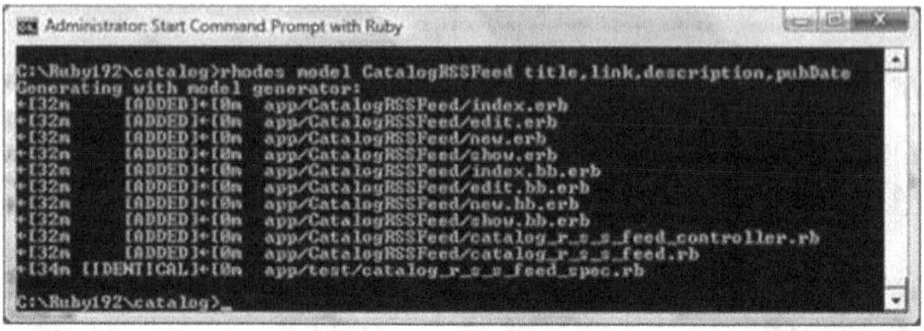

Fig. 1.19 Creating a model for a Rhodes RSS application

Table 1.1 Parameters for the get method

Parameter	Description
:url	URL to send the requests to
:headers	Hash of headers to send with the request
:callback	Callback action to execute when the request is completed
:callback_params	Callback parameters (optional)

```
C:\Ruby192\catalog>rhodes    model    CatalogRSSFeed-
title,link,description,pubDate
```

The model, controller, and view template files get generated in the app/ CatalogRSSFeed folder as shown in Fig. 1.19.

We won't be using the default view templates and actions for CRUD operations, but shall modify the controller class to get the RSS feed, parse the XML feed and display the results in the Android. For XML feed we shall require an XML parser. Rhodes includes the RhoXML parser, which is a lightweight parser and does not support some features. We shall used the ReXML parser, for which add support in the catalog/build.yml file.

```
extensions: ["json", "rexml", "set"]
```

We shall use the AsyncHttp API to get the RSS feed. Use the get(:url, :headers, :callback, :callback_params) method for an HTTP GET request. The parameters for the get method are discussed in Table 1.1.

Send a HTTP request to the IBM developerWorks RSS feed.

```
url =
'http://www.ibm.com/developerworks/views/opensource/rss
/libraryview.jsp'
Rho::AsyncHttp.get(
        :url =>url,
        :callback => (url_for :action =>
        :httpget_callback),
    :callback_param => "" )
```

In the callback method if status is 'ok' get the result of the request.

```
@@get_result = @params['body']
```

Create a REXML::Document object from the result.

```
doc = REXML::Document.new(@@get_result)
```

Using the REXML::XPATH class iterate over the//rss//item elements in the RSS feed and create a CatalogRSSFeed object corresponding to each <item> element.

```
REXML::XPath.each(doc,"//rss//item/") do |e|
CatalogRSSFeed.create(:title
=>e.elements['title'].text,
:link =>e.elements['link'].text,
:description =>e.elements['description'].text,
:pubDate =>e.elements['pubDate'].text)
   end
```

In the index action create an instance variable for all feed results.

```
@catalogrssfeeds = CatalogRSSFeed.find(:all)
```

In the index.erb view template iterate over the @catalogrssfeeds instance variable, which contains the feed results and output the feed titles. A request may be cancelled with the Rho::AsyncHttp.cancel method. The controller file catalog_r_s_s_feed_controller.erb is listed below.

```
require 'rho/rhocontroller'
require 'helpers/browser_helper'

classCatalogRSSFeedController< Rho::RhoController
includeBrowserHelper

def index
    @catalogrssfeeds = CatalogRSSFeed.find(:all)
if @catalogrssfeeds.empty? then
self.update
else


render :action => :index, :back => :exit
end
end


def refresh
CatalogRSSFeed.delete_all
redirect :action => :update
end

def update
url =
'http://www.ibm.com/developerworks/views/opensource/
 rss/libraryview.jsp'

    Rho::AsyncHttp.get(
:url =>url,
:callback => (url_for :action => :httpget_callback),
:callback_param => "" )
render :action => :wait, :back => :exit
end
def show
    @catalogrssfeed = Cata-
logRSSFeed.find(@params['id'])
if @catalogrssfeed
        render :action => :show, :back =>url_for(:action
=> :index )
else
redirect :action => :index
end
end
```

```
defhttpget_callback
if @params['status'] != 'ok'
        @error_params = @params
WebView.navigate( url_for :action => :show_error )
else
        @@get_result = @params['body']
begin
require 'rexml/document'
doc = REXML::Document.new(@@get_result)
REXML::XPath.each(doc,"//rss//item/") do |e|


CatalogRSSFeed.create(:title
=>e.elements['title'].text,
:link =>e.elements['link'].text,
:description =>e.elements['description'].text,
:pubDate =>e.elements['pubDate'].text)
end
            @catalogrssfeeds = Cata-
logRSSFeed.find(:all)
if @catalogrssfeeds.empty?
WebView.navigate( url_for :action => :show_error )
else
WebView.navigate( url_for :action => :index )
end
rescue Exception => e
puts "Error: #{e}"
            @@get_result = "Error: #{e}"
end
end
end
defcancel_httpcall
    Rho::AsyncHttp.cancel( url_for( :action =>
:httpget_callback) )
    @@get_result  = 'Request was cancelled.'
render :action => :index, :back => :exit
end

defget_res
    @@get_result
end
```

```
defget_error
    @@error_params
end
defshow_error
    render :action =>:error, :back =>url_for(:action
=> :index )
end

def exit
    Rho::RhoApplication.close
System.exit
end
  end
```

The index.erb view template is listed below.

```
<div class="pageTitle">
<h1>CatalogRSSFeeds</h1>
</div>
<div class="toolbar">

<div class="regularButton">
<a class="button" href="<%= url_for :action =>:refresh
%>">Refresh</a>
</div>
</div>

<div class="content">
<ul>
<% @catalogrssfeeds.eachdo |catalogrssfeed| %>

<li>
<a href="<%= url_for :action => :show, :id
=>catalogrssfeed.object %>">
<span class="title"><%= catalogrssfeed.title
%></span><span
class="disclosure_indicator"></span>
</a>
</li>

<% end %>
</ul>
  </div>
```

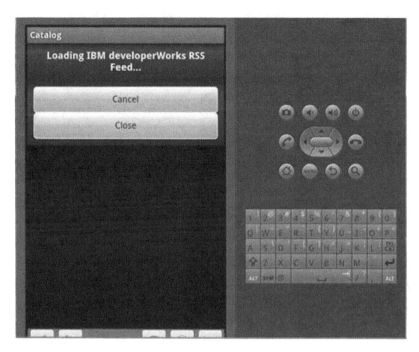

Fig. 1.20 Starting the Rhodes application, the wait message

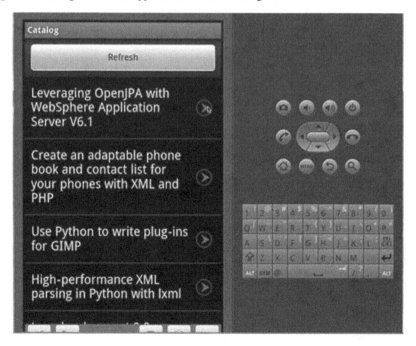

Fig. 1.21 The RSS feed

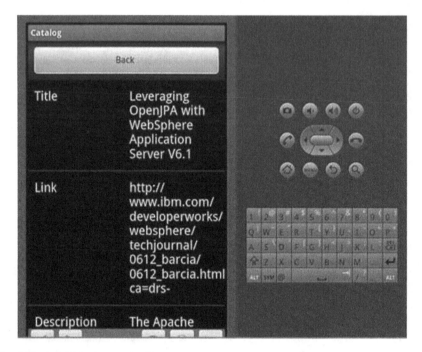

Fig. 1.22 A RSS feed entry detail

Modify the start path in the `catalog/rhoconfig.txt` file for the RSS feed application.

```
# startup page for your application
start_path = '/app/CatalogRSSFeed'
```

Start the Android emulator as before, with the command:

```
C:\Ruby192\catalog>rake run:android
```

The IBM developerWorks RSS Feed application gets started in the Android emulator as shown in Fig. 1.20.

The IBM developerWorks RSS Feed gets listed in Android as shown in Fig. 1.21. Select a feed entry to display the entry detail.

The feed entry detail gets displayed as shown in Fig. 1.22.

The RSS feed may be scrolled to display all the entries. In Chap. 2 we shall discuss developing the same Rhodes application on BlackBerry.

Chapter 2
Rhodes on BlackBerry® Smartphones

BlackBerry® has more than 30% (ranked 2nd) of the smartphone market share. In this chapter we shall, first, introduce using Rhodes on Blackberry JDE, and subsequently develop a Rhodes application to get RSS feed for a magazine on Blackberry JDE. Rhodes uses XRuby to generate the Java code from Ruby code. XRuby compiles Ruby classes to Java class files. Though the Rhodes application is a Ruby application, actually the BlackBerry JDE runs Java, which has been compiled from Ruby. The Ruby and Rhodes installation and configuration procedure is the same as in Chap. 1, but is discussed in this chapter for completeness. The Rhodes application is also the same as in Chap. 1, but is discussed in the context of BlackBerry.

2.1 The ReXML Parser

The ReXML parser provides various classes to parse and process an XML document. Some of those classes are discussed in Table 2.1.

We shall be using only the `Document` and `XPath` classes in this article. Some of the methods in the `Document` class are discussed in Table 2.2.

The `XPath` class provides the methods discussed in Table 2.3.

2.2 Installing the BlackBerry JDE

As a pre-requisite to installing BlackBerry on Windows, install the DirectX SDK from http://www.microsoft.com/download/en/details.aspx?displaylang=en&id=6812. Download the BlackBerry Java Development Environment (JDE) v6.0 from http://us.blackberry.com/developers/javaappdev/javadevenv.jsp. Double-click on the `BlackBerry_JDE_6.0.0.exe` to install the JDE. Install BlackBerry JDE

D. Vohra, *Rhodes Framework for Android™ Platform and BlackBerry® Smartphones,* 23
SpringerBriefs in Computer Science, DOI: 10.1007/978-1-4614-3579-2_2,
© The Author(s) 2012

Table 2.1 ReXML parser

Class	Description
REXML::Attribute	Represents an Element Attribute
REXML::DocType	Represents an XML DOCTYPE declaration
REXML::Document	Represents a full XML document
REXML::Element	Represents an XML Element
REXML::Node	Represents a node
REXML::Parsers:: PullParser	Represents a pull parser
REXML::Parsers:: SAX2Parser	Represents a SAX2 parser
REXML::Text	Represents a text node
REXML::XPath	Wrapper class for XPath functions

Table 2.2 Document class methods

Method	Description
Add	Adds a node
Add_element	Adds an element
Doctype	Returns the DocType of the document if present, or nil
Encoding	Returns the encoding if set, or returns the default encoding
New	Constructor for a new document
Root	Returns the root element
Version	Returns the version if set, or the default version
Write	Outputs the XML document tree
Xml_decl	Returns the XML declaration if set, or the default declaration

Table 2.3 XPath class methods

Method	Description
Each(element, path = nil, namespaces = nil, variables = {})	Takes a context element, the xpath to search for, and a Hash for namespace mapping as parameters, and iterates over nodes that match the specified path. If the xpath is not specified the default xpath is "*"
First(element, path = nil, namespaces = nil, variables = {})	Returns the first nodes that matches the specified xpath. The parameters are the same as the each method
Match(element, path = nil, namespaces = nil, variables = {})	Returns an array of nodes that match the specified xpath

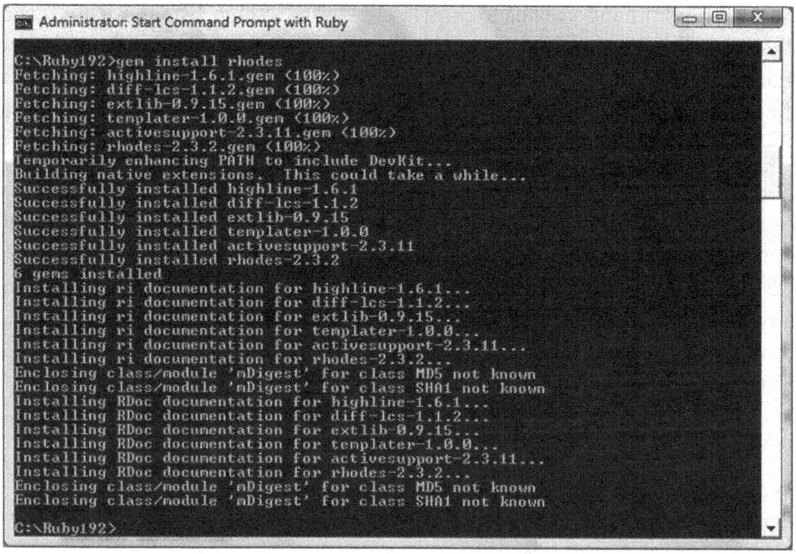

Fig. 2.1 Installing Rhodes

in a directory without spaces in the directory path; for example not in a `C:/Program Files/`sub-directory. The same applies for the JDK 6, which is required for the BlackBerry JDE; install the JDK in a directory without spaces in the directory path. Add `JDK_HOME` and `JDK_HOME/bin` to the `PATH` environment variable. In a later section we shall configure a Rhodes application to used the BlackBerry simulator.

2.3 Installing Rhodes

As Rhodes is a Ruby gem, we need to install Ruby first. As in Chap. 1, download the `rubyinstaller-1.9.2-p180.exe` application. Double-click on the `.exe` file to install Ruby. Install Ruby in a directory without spaces in the directory path as shown in Fig. 2.1. Select the checkbox Add Ruby executables to your `PATH`.

Next, install the Ruby Installer Development Kit (DevKit), which makes it easy to build native Ruby extensions. Download the `DevKit-tdm-32-4.5.1-20101214-1400-sfx.exe`application. Double-click on the `.exe` file and install the self-extracting executables in a directory without spaces, such as `C:/Ruby192/DevKit`. Cd (change directory) to the `DevKit` directory and run the following commands:

```
rubydk.rbinit
rubydk.rb install
```

The init command creates a config.xml file, which lists the RubyInstaller installed Rubies. Install RubyGems, a Ruby packaging system. Download the RubyGems zip file and extract the zip file to a directory, Cd to the directory and run the following command:

```
C:Ruby192\rubygems-1.6.2>ruby setup.rb
```

We also need to install gnuwin32, which provides win32 ports of GNU tools, GNU being a UNIX-like operating system. Download the GetGnuWin32-0.6.3.exe application from http://sourceforge.net/projects/getgnuwin32/files/ getgnuwin32/and double-click on the exe file. Install in a folder (C:/Ruby192 for example). Cd to the installed folder (C:/Ruby192/GetGnuWin32) and run the download command.

```
C:\Ruby192\GetGnuWin32>download
```

Subsequently, run the install command:

```
C:\Ruby192\GetGnuWin32>install C:/gnuwin32
```

Add C:/gnuwin32/bin to the PATH environment variable. Next, install Rake, a Ruby build program, with the following command:

```
C:\Ruby192>gem install rake
```

To install Rhodes run the following command:

```
C:\Ruby192>gem install Rhodes
```

Rhodes 2.3.2 and related gems get installed as shown in Fig. 2.1.

2.4 Creating a Rhodes Application

Rhodes provides an application generator to generate an application. The Rhodes application generator is called rhodes and is run with the following command format:

```
rhodes app <application_name>
```

Before we may run the rhodes command we need to setup Rhodes using the rhodes-setup command as shown in Fig. 2.2. Select Enter for each of the questions. The JDK path should not include any spaces in the directory path. By default BlackBerry JDE version 4.6 or less configuration is checked.

As we are running JDE 6, modify the C:\Ruby192\lib\ruby\gems \1.9.1\gems\rhodes-2.3.2\rhobuild.yml configuration file, listed below, to include the v6.0. The BlackBerry related settings are shown in bold.

```
env:
  app: C:/rhodes-app
  paths:
    java: C:/JDK/Java/jdk1.6.0_24/bin
    android:
    android-ndk:
    cabwiz:
6.0:
    jde:  C:/BlackBerry
    mds:  C:/BlackBerry/MDS
    sim:  9800
  4.6:
    jde:
    mds:
    sim: 9000
  4.2:
    jde:
    mds:
    sim: 8100
build:
  bbpath: platform/bb

  wmpath: platform/wm
  androidpath: platform/android
  iphonepath: platform/iphone
  symbianpath: platform/symbian
  bb:
  bbsignpwd: somepasswordhere
android:
excludedirs:
  all:
  - "**/.*.swo"
  - "**/.*.swn"
  - "**/.DS_Store"
  bb:
  - public/js/iui
  - public/js/jquery*
  - public/jqtouch*
  - public/js/prototype*
  - public/css/iphone*
  - public/iwebkit
  - public/themes
  - "**/jquery*.js"
  - "**/*.db"
```

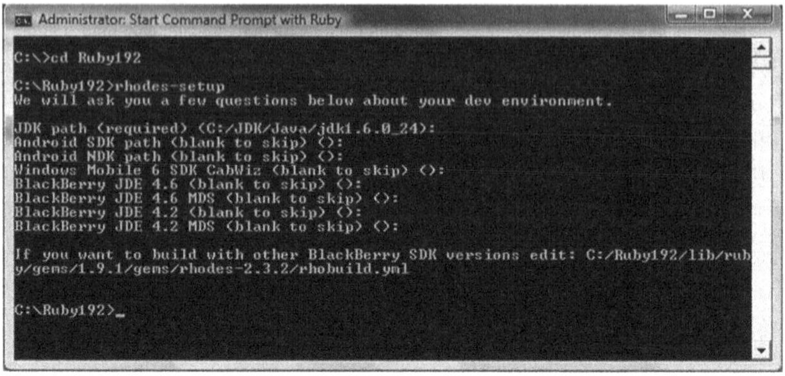

Fig. 2.2 Setting up Rhodes

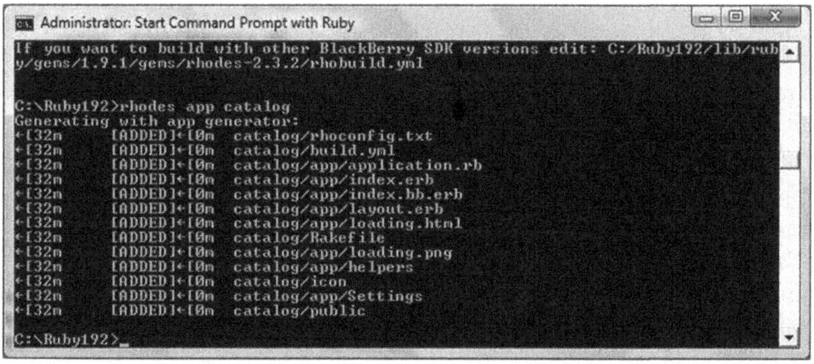

Fig. 2.3 Generating Rhodes application

The jde parameter specifies the directory in which the JDE is installed. The mds parameter specifies the directory in which the BlackBerry Mobile Data Service (MDS) is installed. The simulator port is specified with the sim parameter. Next, run the rhodes application generator to create an application called catalog with the command:

```
>rhodes app catalog
```

The application files get generated in the catalog (application name) directory as shown in Fig. 2.3.

Modify the catalog/build.yml file to specify the BlackBerry version as 6.0. The BlackBerry version is specified with the bbver property.

Fig. 2.4 Compiling and uploading the Rhodes application to BlackBerry JDE

```
sdk:        "C:/Ruby192/lib/ruby/gems/1.9.1/gems/rhodes-
2.3.2"
sdkversion: 2.3.2
name: catalog
version: 1.0
vendor: rhomobile
build: debug
bbver: 6.0
wmsdk: "Windows Mobile 6 Professional SDK (ARMV4I)"
applog: rholog.txt
```

Before developing the application further test the BlackBerry emulator. Cd to
the catalog directory and run the following command.

Fig. 2.5 Rhodes catalog application on BlackBerry

```
C:\Ruby192\catalog>rake run:bb
```

The Rhodes application gets built and uploaded to the BlackBerry simulator as shown in Fig. 2.4.

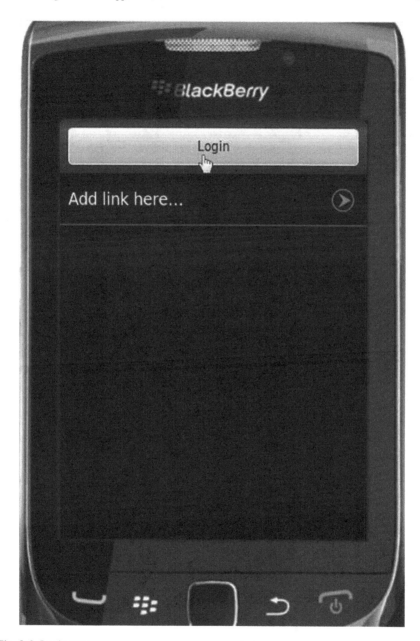

Fig. 2.6 Login page

The BlackBerry simulator gets started. The Rhodes application `catalog` is shown installed on the simulator in Fig. 2.5.

Click on the default Login button as shown in Fig. 2.6.

Fig. 2.7 Testing login

Specify Login and Password and click on Login. The Login page is just a test Login as shown in Fig. 2.7.

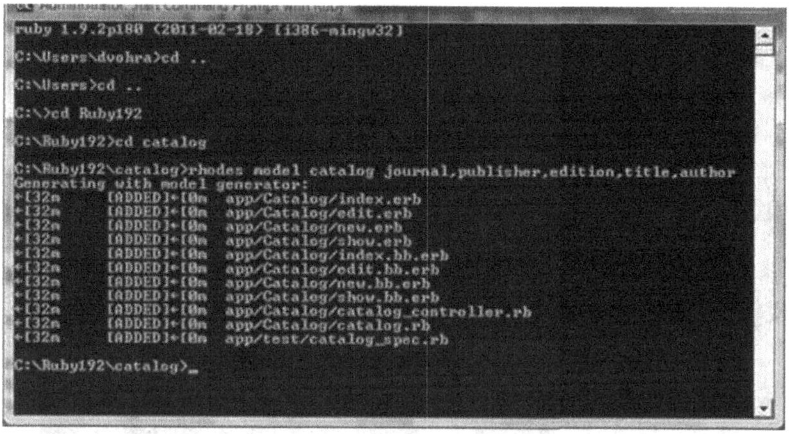

Fig. 2.8 Generating Rhodes model and controller

2.5 Creating a Rhodes Model for a Catalog

Rhodes provides the `rhodes model` command to generate model and controller files, and view templates, and is run with the following command format.

```
rhodes model modelname options
```

The `rhodes model` command generates a scaffolding similar to the Ruby on Rails framework to perform CRUD operations on the model. Cd to the application folder and generate a scaffolding for a catalog, which includes the `journal`, `publisher, edition, title, author` attributes.

```
C:\Ruby192\catalog>rhodes    model    catalog    jour-
nal,publisher,edition,title,author
```

The view templates `index.erb`, `edit.erb`, `new.erb` and `show.erb` get generated as shown in Fig. 2.8. View templates customized for the BlackBerry get generated as `.bb.erb` extension files; `index.bb.erb`, `edit.bb.erb`, `new.bb.erb`, and `show.bb.erb`. Controller file `catalog_controller.rb` and model file `catalog.rb` also get generated.

The controller class extends the `Rho:RhoController` class and includes actions `index`, `edit`, `show`, `new`, `create`, `update` and `delete` for CRUD operations.

```
class CatalogController < Rho::RhoController

end
```

Next, we shall upload the application to the BlackBerry emulator and test the application in the emulator. Before we may run the emulator we need to specify the application to run in the `catalog/rhoconfig.txt` file.

Fig. 2.9 Compiling and uploading Rhodes application to BlackBerry

```
# startup page for your application
start_path = '/app/Catalog'
```

To run the emulator and upload the rhodes model catalog run the command:

```
C:\Ruby192\catalog>rake run:bb
```

The BlackBerry simulator gets started. The Rhodes application gets compiled to Java code. The Java code compiled from Ruby code gets packaged into catalog.jar and gets uploaded to the BlackBerry simulator (Fig. 2.9).

The catalog application gets started in the emulator. Click on **New** to create a catalog entry as shown in Fig. 2.10.

Specify **Journal**, **Publisher**, **Edition**, **Title** and **Author** and click on **Create** as shown in Fig. 2.11.

Similarly, new catalog entries may be added. Click on the icon for a catalog entry to display the entry as shown in Fig. 2.12.

The selected catalog entry gets listed as shown in Fig. 2.13.

2.6 Creating a Rhodes Model to get RSS Feed

In the previous sections we have only tested the default model generated by Rhodes. In this section we shall create a rhodes model to get a RSS feed, which is essentially an XML document, parse the XML document, and display the feed in the

Fig. 2.10   Creating a new catalog entry

BlackBerry. We shall use the Oracle Magazine RSS Feed (http://www.oracle. com/ocom/groups/public/@otn/documents/webcontent/rss-oramag-recent.xml) for the example. RSS Feed is essentially an XML document, which we shall parse using

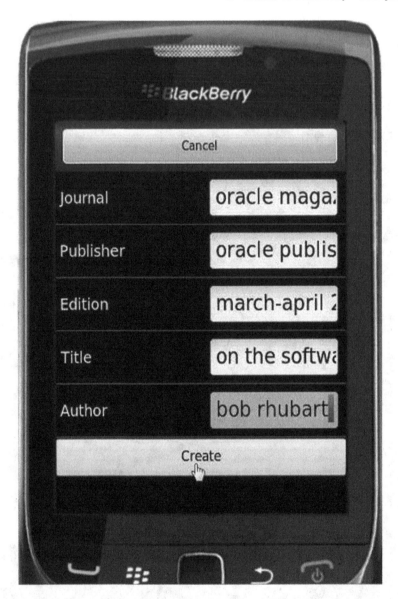

Fig. 2.11 Specifying catalog entry attributes

the ReXML parser. The RSS feed is in XML format and contains entries as
`<item></item>` elements, the root element being `<rss></rss>`. A section
of the RSS Feed XML document for the Oracle Magazine is listed below.

```
<?xml version="1.0" encoding="UTF-8"?>
<rss version="2.0">
<channel>
<title>Oracle Magazine - Most Recent</title>
<link>http://www.oracle.com/technology/oramag/oracle</l
ink>
<description>Here are the latest Oracle Magazine ar-
ticles, columns, and issues.
</description>
<language>en-us</language>
<copyright>Copyright 2008 Oracle. All Rights Re-
served.</copyright>

<managingEditor>opubedit_us@oracle.com</managingEditor>
<pubDate>Mon, 19 Dec 2005 22:04:11 GMT</pubDate>
<lastBuildDate>Fri, 9 Sep 2011 22:21:14
GMT</lastBuildDate>
<item>
<title>Architect: Getting Schooled</title>
<link>http://www.oracle.com/technetwork/issue-
archive/2011/11-sep/o51
     architect-445768.html</link>
<description>Education, training, and experience are
stepping stones to a career as
     a software architect.</description>

<guid isPermaLink="false">{89f1fb2-f946-6951-5b7a-
5f5f54ee330}</guid>
<pubDate>Fri, 9 Sep 2011 22:21:14 GMT</pubDate>
</item>
<item>
<title>PL/SQL: Working with Strings</title>
<link>http://www.oracle.com/technetwork/issue-
archive/2011/11-sep/o51plsql-453456.
     html</link>
<description>Part 3 in a series of articles on under-
standing and using PL/SQL
</description>

<guid isPermaLink="false">{89f1fb2-f946-6951-5b7a-
5f5f54ee330}</guid>
<pubDate>Tue, 6 Sep 2011 20:34:27 GMT</pubDate>
</item>
...
...
</channel></rss>
```

Fig. 2.12 New catalog entry

Each item element has sub-elements title, link, description, guid, and pubDate. We shall be parsing these sub-elements and displaying their values using a Rhodes model. Create a Rhodes model CatalogRSSFeed with attributes title, link, description, and date with the following command:

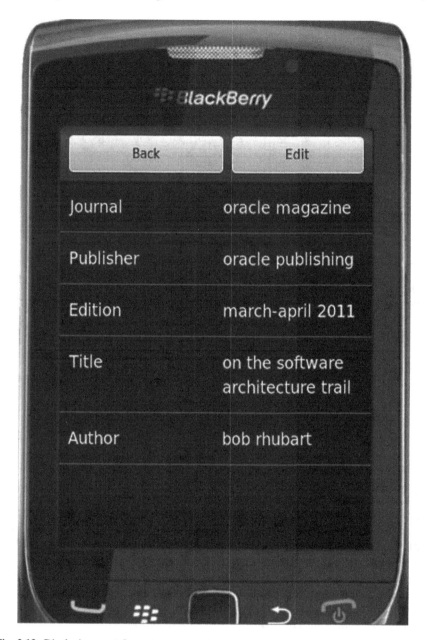

Fig. 2.13 Displaying a catalog entry

```
C:\Ruby192\catalog>Rhodes      model      CatalogRSSFeed
title,link,description,pubDate
```

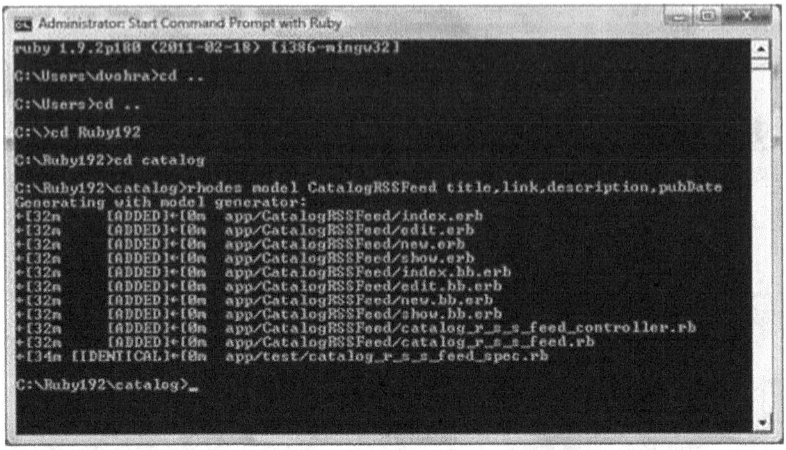

Fig. 2.14 Creating a RSS feed application with Rhodes

Table 2.4 GET method request parameters

Parameter	Description
:url	URL to send the requets to
:headers	Hash of headers to send with the request
:callback	Callback action to execute when the request is completed
:callback_params	Callback parameters (optional)
:authentication	Sends basic Auth header with the request (optional)
:ssl_verify_peer	Verifies SSL certificates (optional). True by default

The model, controller, and view template files get generated in the app/CatalogRSSFeed folder as shown in Fig. 2.14.

The model class CatalogRSSFeed extends the Rhom:RhomObject class. We won't be using the default view templates and controller actions for CRUD operations, but shall modify the controller class to get the RSS feed, parse the XML feed and display the results in the BlackBerry. For XML feed we shall require an XML parser. Rhodes includes the RhoXML parser, which is a light-weight parser and does not support some features. We shall used the ReXML parser, for which add support in the catalog/build.yml file

```
extensions: ["json", "rexml", "set"]
```

We shall use the AsyncHttp API to get the RSS feed. Use the get(:url, :headers, :callback, :callback_params) method for an HTTPGET request. The parameters for the get method are discussed in Table 2.4.

Specify the url to the Oracle Magazine RSS Feed. Send a HTTP request to the RSS feed using the Rho::AsyncHttp.get method.

Table 2.5 AsychHttp
callback parameters

Parameter	Description
@params["body"]	The body of the Http response
@params["headers"]	The response headers hash
@params["cookies"]	The server cookies
@params["http_error"]	The HTTP error code, if response code is not 200

```
url                                                         =
'http://www.oracle.com/ocom/groups/public/@otn/document
s/webcontent/rss-oramag-recent.xml'
  Rho::AsyncHttp.get(
            :url => url,
            :callback =>
            (url_for :action => :httpget_callback),
            :callback_param => "" )
```

The AsychHttp callback has the following parameters, listed in Table 2.5, available.
In the callback method if status is 'ok' get the result of the request.

```
@@get_result = @params['body']
```

Create a REXML::Document object from the result using the new constructor.

```
doc = REXML::Document.new(@@get_result)
```

Using the REXML::XPath class iterate over the//rss//item elements in the
RSS feed using the REXML::XPath:each method, which returns an array of
nodes, and create a CatalogRSSFeed object corresponding to each node using
the create method of the model.

```
REXML::XPath.each(doc,"//rss//item/") do |e|
    CatalogRSSFeed.create(:title                           =>
e.elements['title'].text,
        :link => e.elements['link'].text,
   :description => e.elements['description'].text,
      :pubDate => e.elements['pubDate'].text)
    end
```

In the index action create an instance variable for all feed results using the
find(:all) method.

```
@catalogrssfeeds = CatalogRSSFeed.find(:all)
```

In the index.bb.erb view template iterate over the @catalogrssfeeds
instance variable, which contains the feed results and display the feed titles with a
link to the RSS feed entry detail using the show.bb.erb view template.

```
<% @catalogrssfeeds.each do |obj| %>
  <td><%= link_to "#{obj.title}", :action => :show,
:id =>
    obj.object %></td>
  <% end %>
```

The show.bb.erb view template shows the RSS feed for an entry and displays the title, link, description, and publication date. The show.bb.erb view template is listed below.

```
<div id="pageTitle">
 <h1><%= @catalogrssfeed.title%></h1>
</div>
<div id="toolbar">
<%= link_to "Back", :action => :index %>
<%= link_to "Edit", :action => :edit, :id =>
@catalogrssfeed.object %>
</div>
<div id="content">
<table>
<tr>
 <td class="itemLabel">Title</td>
 <td                         class="itemValue"><%=
@catalogrssfeed.title%></td>
 </tr>
 <tr>
 <td class="itemLabel">Link</td>
 <td class="itemValue"><%= @catalogrssfeed.link%></td>
 </tr>
 <tr>
 <td class="itemLabel">Description</td>
 <td                         class="itemValue"><%=
@catalogrssfeed.description%></td>
 </tr>
 <tr>
 <td class="itemLabel">PubDate</td>
 <td                         class="itemValue"><%=
@catalogrssfeed.pubDate%></td>
 </tr>
 </table>
 </div>
```

A request may be cancelled with the Rho::AsyncHttp.cancel method. The controller file catalog_r_s_s_feed_controller.erb is listed below.

```
require 'rho/rhocontroller'
require 'helpers/browser_helper'

class CatalogRSSFeedController < Rho::RhoController
  include BrowserHelper
def index
    @catalogrssfeeds = CatalogRSSFeed.find(:all)
if @catalogrssfeeds.empty? then
  self.update
else
render :action => :index, :back => :exit
 end
   end

def refresh
   CatalogRSSFeed.delete_all
   redirect :action => :update
end

def update
      url                                           =
'http://www.oracle.com/ocom/groups/public/@otn/document
s/webcontent/rss-oramag-recent.xml'
    Rho::AsyncHttp.get(
      :url => url,
      :callback    =>    (url_for    :action    =>
:httpget_callback),
       :callback_param => "" )
   render :action => :wait, :back => :exit
  end
  def show
    @catalogrssfeed                =                Cata-
logRSSFeed.find(@params['id'])
     if @catalogrssfeed
       render :action => :show, :back => url_for(
:action => :index )
     else
       redirect :action => :index
     end
   end
 def httpget_callback
    if @params['status'] != 'ok'
       @error_params = @params
       WebView.navigate  (  url_for  :action  =>
:show_error )
     else
```

```
        @@get_result = @params['body']
        begin
            require 'rexml/document'

  doc = REXML::Document.new(@@get_result)
  REXML::XPath.each(doc,"//rss//item/") do |e|
    CatalogRSSFeed.create(:title                    =>
e.elements['title'].text,
        :link => e.elements['link'].text,
    :description => e.elements['description'].text,
        :pubDate => e.elements['pubDate'].text)
    end
            @catalogrssfeeds          =          Cata-
logRSSFeed.find(:all)
            if @catalogrssfeeds.empty?
                WebView.navigate ( url_for :action =>
:show_error )
            else
                WebView.navigate ( url_for :action =>
:index )
            end
        rescue Exception => e
            puts "Error: #{e}"
            @@get_result = "Error: #{e}"
        end
    end
    end

    def cancel_httpcall
      Rho::AsyncHttp.cancel(   url_for(   :action   =>
:httpget_callback) )
        @@get_result = 'Request was cancelled.'
        render :action => :index, :back => :exit
    end

  def get_res
    @@get_result
  end

  def get_error
    @@error_params
  end
  def show_error
      render :action  => :error,  :back  => url_for(
:action => :index )
    end
```

```
def exit
  Rho::RhoApplication.close
  System.exit
end
end
```

The index.erb view template is listed below.

```
<div class="pageTitle">
<h1>CatalogRSSFeeds</h1>
</div>

<div class="toolbar">

<div class="regularButton">
<a    class="button"    href="<%=    url_for    :action    =>
:refresh %>">Refresh</a>
</div>
</div>

<div class="content">
<ul>
<% @catalogrssfeeds.each do |catalogrssfeed| %>

<li>
<a href="<%= url_for :action => :show, :id => cata-
logrssfeed.object %>">
<span    class="title"><%=    catalogrssfeed.title
%></span><span class="disclosure_indicator"></span>
</a>
</li>

<% end %>
</ul>
</div>
```

Modify the start path in the catalog/rhoconfig.txt file for the RSS feed application.

```
# startup page for your application
start_path = '/app/CatalogRSSFeed'
```

Start the BlackBerry emulator as before, with the command:

```
C:\Ruby192\catalog>rake run:bb
```

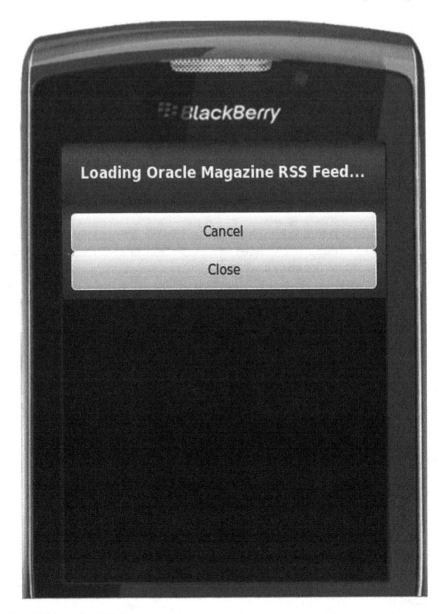

Fig. 2.15 Loading RSS feed

The Oracle Magazine RSS Feed application gets started in the BlackBerry emulator as shown in Fig. 2.15.

The Oracle magazine RSS feed gets listed in BlackBerry as shown in Fig. 2.16. Select a feed entry to display the entry detail.

Fig. 2.16   Getting RSS feed with Rhodes on BlackBerry

The feed entry gets displayed as shown in Fig. 2.17.
The RSS feed may be scrolled to display all the entries as shown in Fig. 2.18.

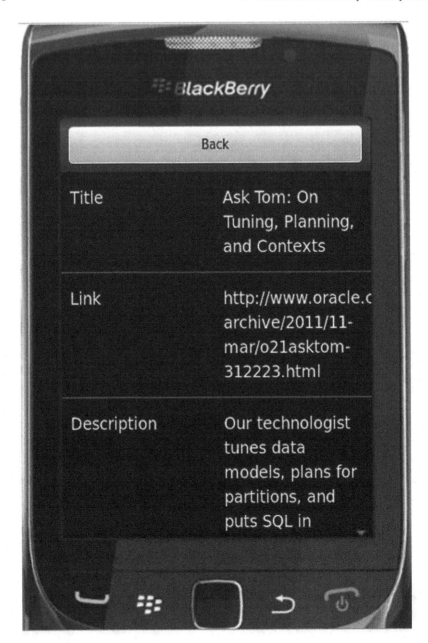

Fig. 2.17 Displaying a RSS feed entry

In this chapter we discussed using Rhodes with BlackBerry. In the previous chapter we discussed Rhodes with Android. Rhodes with BlackBerry has the following differences from Rhodes with Android.

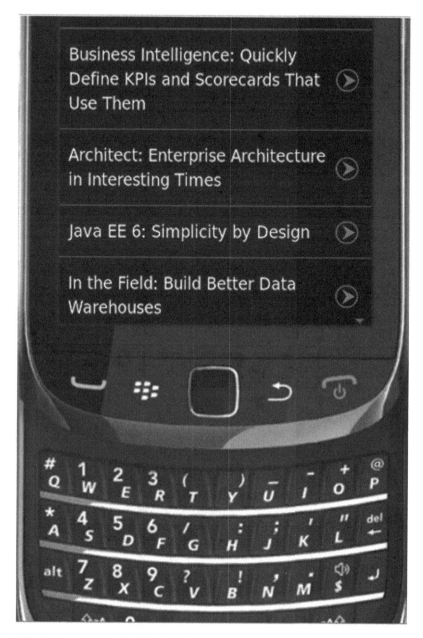

Fig. 2.18 Scrolled view of RSS feed on BlackBerry

- The view templates used are different. The `*.bb.erb` view templates are used with BlackBerry instead of the `*.erb` view templates with Android.
- The BlackBerry emulator is different from the Android emulator, and the command to run the emulator is different.

- The configuration with BlackBerry is different.
- The required software is different with BlackBerry. The BlackBerry JDE and DirectX SDK are used with BlackBerry instead of Eclipse and ADT with Android.